AGRICULTURE
in North Carolina Before the Civil War

Cornelius O. Cathey

Division of Archives and History
NORTH CAROLINA DEPARTMENT OF CULTURAL RESOURCES
Raleigh
Second Printing
1974

Department of Cultural Resources
Mrs. Grace J. Rohrer, *Secretary*

Division of Archives and History
Robert E. Stipe, *Director*

North Carolina Historical Commission
T. Harry Gatton, *Chairman*

Miss Gertrude Sprague Carraway
Gordon S. Dugger
Frontis W. Johnston

J. C. Knowles
Hugh T. Lefler
Edward W. Phifer, Jr.

CONTENTS

Geographical Influences	1
Earliest Settlements and Settlers	2
Land Grants	3
Agricultural Experimentation in the Colonial Period	4
Production of Food Supplies	4
Corn	5
Other Grains	5
Rice	6
Vegetables	6
Fruits	8
Cotton Production	8
Tobacco as a Commercial Crop	9
Livestock Production	10
Beekeeping and Honey Production	12
Over-all Aspects of Colonial Agriculture	12
Effects of the Revolution	13
Conditions Following the Revolution	15
Westward Migration	17
Agricultural Revolution	18
Publicity for Agricultural Reform	21
Agricultural Societies	23
Renewed Interest in Reform, 1845-1860	24
Organization of a New Agricultural Society	25
"Scientific" Farming	28
Fertilizers	29
Farm Implements	29
Commercial and Subsistence Crops	31
Cotton	33
Tobacco	34
Corn	36
Wheat	37
Rice	37
Fruit	39
Vegetables	40
Livestock	42
Farms in 1860	44

LIST OF ILLUSTRATIONS

A farmer using an 1837 cultivator .... *Frontispiece*

Indian harvest dance 2

Elkanah Watson 7

Pig scalding in preparation of meat 11

Model of the cotton gin 14

Archibald D. Murphey 15

Notice of sale of farm and household property 17

Bill of sale of farm products, 1807 19

Silver medal awarded by North Carolina Agricultural Society 26

Thomas Ruffin 27

Wheel cultivator, patented in 1846 30

Cotton growth and harvest 32

Transportation of cotton to market 32

Use of the saw gin 32

Threshing wheat by use of hand flails 38

Threshing wheat with horses 38

Livestock feeder 43

Preface

North Carolina has long been recognized as a leading agricultural state, but in the years preceding the Civil War she was not known for her large plantations as were some of the southern states. Though there were some planters and some slaveholders, by far the majority of North Carolina's farmers worked their few acres themselves and struggled to provide the necessities of life for their families.

This pamphlet describes farming conditions and methods in the colonial period and in the years following the Revolution and preceding the Civil War. Dr. C. O. Cathey, the author, was professor of history at the University of North Carolina at Chapel Hill until his recent retirement. For a number of years he also served as dean of student affairs at the university. He is well known for his book, *Agricultural Developments in North Carolina, 1783-1860*, published in 1956 by the University of North Carolina Press.

Illustrations are from the files of the Division of Archives and History and from Benjamin Butterworth (arranger and compiler), *The Growth of Industrial Art* (Washington: Government Printing Office, 1892). Copies used here were reproduced by Mrs. Madlin Futrell, a member of the staff of the Museums Section, and by Messrs. Bob Allen and Gene Furr, former staff members. Mrs. Ann W. Little, who formerly worked in the Historical Publications Section, assisted in seeing the pamphlet through the press.

Memory F. Mitchell
Historical Publications Administrator

October 1, 1974

A farmer using an 1837 cultivator. From Butterworth, *The Growth of Industrial Art*.

AGRICULTURE
in North Carolina Before the Civil War

Geographical Influences

North Carolina encompasses 52,712 square miles of area, is 503 miles long, averages about 100 miles in width, and has at its widest point about 300 miles of frontage on the Atlantic Ocean. Three well-defined geographic regions are to be found in the state: the Atlantic Coastal Plain which comprises about one-half of the area of the state and extends from the coast westward to the fall line of the rivers, the Piedmont which extends from the fall line on the East to the Appalachians in the West, and the Mountain Region. Differences of soil, elevation, rainfall, and length of growing season between these regions combine to give North Carolina an ability to produce, with some degree of success, practically every crop grown in the temperate zone.

Agriculture has been the principal source of livelihood for the people of the state from the time of the first settlement almost to the present day. Although much of the soil lacked those characteristics that denote top quality, nature had been generous in other respects to the land. The average growing season ranges from a minimum of about 200 days in some parts of the mountains to a maximum of about 240 days in the southeastern part of the state. The average annual rainfall is 50 to

60 inches, and, for agricultural purposes, is well distributed in the normal year. Professor Samuel Huntington Hobbs, Jr., an authority on the economic and social life of the state, wrote:

> Taking all factors into consideration—extremes of heat and cold, length of growing season, amount and regularity of rainfall, clear days, absence of storms, and so on—North Carolina averages up about as well as any area that can be presented. . . . While there are climatic variations, they are not extreme.

Indians, who taught the colonists much about agriculture, doing a harvest dance. From Theodore de Bry's engraving of John White's painting. Files of Department of Archives and History.

Earliest Settlements and Settlers

The first permanent settlements by white men in what later became the colony of Carolina were made by Virginians in the Albemarle Sound area about the middle of the seventeenth century. The desire for new land for agricultural purposes, which accounts to a considerable extent for the rapidity of settlement of the nation, motivated these first settlers. After its creation in 1663, in an effort to speed up settlement of the colony, the government established policies which made it easy for a person to acquire land in Carolina. That part of the colony that later became North Carolina, to stimulate further settlement, fixed 660 acres as the maximum land grant to an individual. Although ways were found to breach that ceiling, the colony of North Carolina for this and other reasons tended from the beginning to be a community of moderate-sized farms rather than of large plantations. Essentially the farmer was a laborer who farmed for a living; whereas, the planter was a businessman who planted for profits. The fact that farming rather than planting was the major emphasis in agriculture must be borne in mind in interpreting social and political developments in pre-Civil War North Carolina.

Handicapped by the lack of a good seaport, the population of North Carolina grew slowly but steadily throughout the Colonial period by migration of settlers principally from Virginia, South Carolina, and Pennsylvania. Most of the people who settled in the Coastal Plain and along the Virginia border were English in origin. From the 1720's onward a tide of migration of Scotch-Irish and Germans began to flow into the Piedmont, and, by the end of the Colonial period, a few settlers had moved through the passes in the mountains into what is today Tennessee and Kentucky. In smaller numbers, Scotch-Highlanders settled in the valley of the Cape Fear River, where, along with the pursuit of agriculture, they established a profitable naval stores industry. A sprinkling of French Huguenots and Jews also settled in the colony—principally in the small towns along the coast. The first official census of North Carolina, that of 1790, indicated that approximately one-fourth of the population was Negro. These people were to be found in all parts of the state, but their heaviest concentration was in those counties of the East and along the Virginia border where, as slaves, they provided the labor force in the tobacco fields.

Land Grants

A new arrival in the colony found little difficulty in securing a patent to the plot of land of his choice. Having made his selection, the settler had a survey made and a plat of the land drawn by the official surveyor. These data were then dispatched to the governor who issued patents to the land in the names of the Lord Proprietors, or, after 1729, in the name of the King.

Generally, the survey of boundaries was carelessly and inaccurately done—a fact that led to considerable dispute and litigation between neighbors. In the Colonial period, the land holder in North Carolina was expected to pay a quitrent—a nominal sum per acre which quitted him of rendering any other feudal-like service to the grantor. These payments ceased upon the outbreak of the American Revolution.

Agricultural Experimentation in the Colonial Period

The Colonial period was one of widespread experimentation in agriculture in North Carolina and in the other English colonies. The settlers, naturally, attempted to grow all the fruits, grains, and vegetables that they had grown in Europe. To these were added such crops obtained from the Indians as corn, tobacco, potatoes, and various types of beans and peas. Also, since North Carolina occupies the same belt of the globe as Gibraltar, Malta, Syria, and Iran—which grow such semitropical products as dates and citrus fruits—unsuccessful attempts were made to produce the same crops in the colony. Repeated efforts to establish a silk industry in North Carolina also failed. In the over-all picture, however, significant accomplishments sprang from the Colonial farmer's experimentation. He soon adopted the crops and planting methods employed by the Indian, transplanted many crops to America that had not previously been grown here, and learned to till the soil extensively rather than intensively. Although many new practices and techniques have been introduced into agriculture since his time, remarkably few crops are now being planted with which he was not familiar. In fact the Colonial North Carolinian planted a few crops, such as indigo and rice, which are not now planted in the state.

Production of Food Supplies

The circumstances under which the Colonial settler lived—remote from his neighbors, and practically without access to market places—necessitated that he give major emphasis to the production of the family food supply. This was the primary objective of all farmers and most of the planters in North Carolina down to the Civil War, and one that was easily attained. The land produced a great variety of subsistence crops, cereals of all sorts that are now being grown in the state, vegetables in abundance, and fruits from orchards and vineyards. The herds of livestock which roamed the fields and forests, along with wild game and fish which abounded in forests and streams, provided an abundant supply of meat. The ease with which a man could make a living in Colonial North Carolina undoubtedly led many settlers to become less efficient in their farming practices and to develop attitudes which served as obstacles to the reform of agriculture at a later date.

Corn

From the very beginning Indian maize, or corn, was the most valuable subsistence crop produced in the colony. Fortunately, corn could be grown in every part of North Carolina, and it soon became the most important single item of food for both man and beast. John Lawson, who wrote the first history of North Carolina, described corn as "the most useful grain in the world." As a crop, corn was particularly well suited to culture in hills—a practice borrowed by the farmer from the Indian, and more widely used than any other method of culture in the Colonial period. In fact, the records left by some of the largest planters in North Carolina indicate that, as late as the Civil War, some were still measuring the size of their crops by the number of hills planted rather than by acres. Generally, the hills were prepared by hand labor, using the hoe or mattock, in fields from which the stumps and roots of trees had not been removed. No effort was made to align the hills in rows for cultivation by animal-drawn implements. Most of the cultivation of corn and other crops was done with hand tools. If available, a fish was dug into the bottom of each hill as a fertilizer. The Colonial farmer also followed the Indian practice of intertilling peas, beans, and squash with the corn. The hills were dug up again and planted with the same crops, year after year, until the soil ceased to be sufficiently productive. Then, the farmer cleared new ground and started the process over again—no effort was made to prevent soil exhaustion or to restore fertility to the soil.

Other Grains

The Coastal Plain, where the first settlements were made, was not well adapted to the culture of the smaller grains such as wheat, oats, barley, millet, buckwheat, and rye; very little of these crops was produced in that part of the colony. The settlers who later flowed into the Piedmont, however, early turned to the cultivation of these cereals. Corn far overshadowed in importance all of the grain crops because it could be grown over the entire colony, required no extraordinary amount of labor to produce, could be harvested at the farmer's leisure, was better adapted to the pioneer's system of culture, and almost never failed to produce a return. Besides, corn was far more widely used as a food by both man and beast.

Rice

The culture of rice was confined to the maritime parts of the colony, particularly the area drained by the lower Cape Fear River. This area marked the northern limits of the "rice coast," which included the entire seaboard of South Carolina and Georgia and in which the culture of rice was the most important agricultural activity. Although the colony of North Carolina did not grow much rice, that which she did produce was considered of very high quality and was much in demand by South Carolina planters for use as seed.

Vegetables

Widespread efforts were made in the Colonial period to grow the vegetables which the settler had been accustomed to raising in Europe or in his place of origin. To these were added vegetables that were native to the New World. Vegetable production, however, did not require much of the farmer's time. The absence of markets, wherein surpluses might be disposed of, the limited knowledge and facilities for preserving vegetables, and the lack of appreciation of what the present-day American calls a "balanced diet" all contributed to a neglect of this phase of agriculture. Meat, which in that day comprised a disproportionately large part of the diet, could be had in great variety and plenty for almost nothing. John Lawson described the Colonial North Carolina gardener as "negligent and unskilful."

The cultivation of "Irish" and sweet potatoes, both natives of the New World, became widespread in North Carolina. The many varieties of peas and beans would, perhaps, follow potatoes as the next important "garden" vegetables produced in the colony.

Elkanah Watson, early advocate of agricultural reform. Engraving by J. W. Paradise, from Collections of the Library of Congress. Files of Department of Archives and History.

In fact, the adaptability of the soils and climate of North Carolina for the production of peas and beans provided the colony with one of its first items of importance in commerce. Elkanah Watson, one of the nation's earliest advocates of agricultural reform, visited a planter near Halifax, in 1778, who was producing around 4,000 bushels of peas a year for shipment to the

West Indies. The value of peas for use in rotation schemes to maintain soil fertility was not known in the Colonial period, and they were not planted for that purpose. Of lesser importance than the above, one might have found some or all of the following vegetables in a Colonial garden: carrots, leeks, parsnips, turnips, artichokes, radishes, beets, onions, shallots, chives, pumpkins, squashes, cucumbers, melons, lettuce, cabbage, spinach, fennel, purslane, samphire, rhubarb, rocket, cress, parsley, asparagus, and coleworts. Tomatoes, or "love apples" as they were called, were grown for ornamentation purposes only.

Fruits

Most of the fruits now produced in North Carolina were first introduced during the Colonial period. Peaches, apples, and grapes particularly were grown in great abundance. By the standards of that day, the quality of these products was considered superior. The unidentified author of *American Husbandry,* published in London in 1775, and regarded as the "most accurate and comprehensive" critic of Colonial American agriculture, said: "Fruit in none of the colonies is in greater plenty [than in North Carolina], or finer flavour; they have every sort that has hitherto been mentioned in this work; peaches as in the central colonies, are so plentiful, that the major part of the crop goes to the hogs." Figs, cherries, apricots, pears, plums, pecans, quinces, damsons, and nectarines were also grown. There was no market for surpluses of fruit, and no means of preserving it except by drying, or making brandies. Under such circumstances, the orchards were usually poorly kept, unfenced, and hogs and cattle foraged in them at will. Despite the above comment, the planting of an orchard was usually a part of the settling process when a new tract of land was occupied; and, undoubtedly, more farmers produced their own fruit then than in the present day. The farmer also made greater use of the great variety of berries and nuts that abounded in the fields and forests of the colony.

Cotton Production

England's interest in woolen and linen manufacturing led her to discourage the production of cotton in her colonies. Cotton was produced, however, in Colonial North Carolina for home consumption. A visitor during the American Revolution observed that cotton was "troublesome" to grow, gather, and separate from the seed, but, if managed properly, "it would be an Article of great consequence." Although the British blockade during the Revolution stimulated cotton growing in the state, the full development of cotton as a commercial crop did not come until new techniques of production, processing, and manufacturing were evolved in a later period.

Tobacco as a Commercial Crop

Tobacco was, by far, the most important commercial crop grown in Colonial North Carolina. Its culture, however, was confined generally to the tier of counties along the Virginia line and the Albemarle Sound area and never became the all-absorbing interest to producers that it did in Colonial Virginia and Maryland. In fact, the area in which tobacco was grown in North Carolina may be considered as but an extension of the Virginia tobacco region. The planters drew heavily on Virginia experience with the crop and marketed their product in Virginia markets where some claimed North Carolina tobacco was discriminated against because it competed with Virginia's "grand staple."

Although North Carolina ranked as a very poor third to Virginia and Maryland as a producer of tobacco, most of the colony's plantation-scale agricultural units were devoted to the production of that crop. The heaviest concentrations of slaves, and the largest landholdings were in the tobacco-growing parts of the colony. The type of leaf grown at that time was found to grow best in freshly cleared land which had accumulated through processes of vegetable decay a surface stratum of rich dark mold. Hills were dug up in this soil in which tobacco was planted year after year until the soil ceased to be sufficiently productive. No effort was expended to preserve the basic fertility of the soil or to prevent its wastage by erosion. As a result, the tobacco planter usually employed his work force during part of the year in clearing new ground for the crop. Under such practice, the tobacco plantation was necessarily large. The prevailing opinion was that the tobacco planter should own fifty acres of land for each member of his working force. As one observer expressed it, "with less than this they will find themselves distressed for want of room." When land suitable for the production of the crop became scarce, the planter sold his old lands for use in the production of corn or wheat, and moved toward the Piedmont section. By the end of the Colonial period, this shift was well under way in North Carolina.

Livestock Production

The early North Carolinian obtained more profit as a producer of livestock, particularly cattle and swine, than from any other agricultural source. By the end of the period the land was only sparsely settled, and the vast unsettled area served the people as a great "commons" upon which their livestock was free to forage at will. The "open range" practice relieved the livestock owner of any obligation to fence his stock in. Rather, it was necessary that the farmer maintain fences around his cultivated fields to keep the stock out. Cattle, sheep, swine, and horses, none native to the New World, were in abundant supply in the colony. There were also a few goats, but no mules. The ox and horse shared in performing draft services. The author of *American Husbandry,* referred to above, in commenting on livestock in Colonial North Carolina, observed that it was "not an uncommon thing to see one man the master of from 300 to 1,200 and even 2,000 cows, bulls, oxen, and young cattle; hogs also in prodigious numbers." The owner of such a herd would have his own peculiar brand and mark, which would be recorded at the county seat, to facilitate identification of his property. Twice a year, usually, the livestock was rounded up into pens or corrals where ownership was determined and the branding done. Owners of land had a property right, called a "woods-right," in the wild or unmarked livestock found in the vicinity of their herds.

Because of the mildness of the weather in North Carolina, the average livestock owner considered it unnecessary to provide his herds with either forage or shelter during the winter. Consequently, very little attention was given the improvement of meadows, or the accumulation of forage for winter feeding. A traveler in the state in 1783 noted that "most of the farmers, although they keep a number of cattle in the woods, can hardly winter one milch cow at the house." Suffered as they were to shift for themselves with practically no care as to their shelter, feeding, and breeding, the quality of livestock tended to deteriorate. Some owners as a result of this neglect, were without milk, butter, and cheese even though possessed of vast numbers of cows. Undoubtedly, the losses resulting from disease, exposure, depredations of other animals, insect pests, and theft were enormous. Swine, by their very nature, were best adapted to

Early version of pig scalding as a step in the preparation of the pig for meat. From Butterworth, *The Growth of Industrial Art.*

survive under those circumstances. "Nowhere on the whole continent," wrote a visitor in 1783, "is the breeding of swine so considerable or so profitable as in North Carolina." In contrast, sheep were the least well-adapted to survive.

Unhampered by present-day standards as to quality, large numbers of livestock on the hoof, and quantities of livestock products, such as salted beef, pork, and butter, found ready acceptance in the channels of Colonial trade. Because of numerous complaints as to methods employed in the colony in processing livestock products, the practice of marketing the product on the hoof was becoming more popular toward the end of the period. By that time, herds of cattle and swine moving along the roads to markets, chiefly in Norfolk, Baltimore, Philadelphia, and New York had become a familiar sight. The keeper of the Bethabara Dairy, in the Moravian community near present-day Winston-Salem, noted on October 20, 1774, that during September and October "more than 1000 head of cattle have been driven by here on the way to Pennsylvania." Upon arrival at their destination, these cattle were usually fattened, slaughtered, and marketed as local products. Thin as they were upon arrival, these animals brought from three to six dollars a head, which, under the prevailing practice in the livestock business, was almost clear profit—except for the expense of the drovers. By the end of the Colonial period, it is estimated that approximately 30,000 cattle, and from 30,000 to 50,000 hogs were being driven to market annually.

Although wild fowl abounded in North Carolina, nearly every settler made some effort to raise his own supply of chickens, ducks, geese, and turkeys. These, like other kinds of livestock, had little provision made for their shelter and feeding, and usually inclined toward a wild state.

Beekeeping and Honey Production

The importance of beekeeping as a source of sweets for the settler's table, or of another item in trade, is reflected by nearly all statistics of production in Colonial North Carolina. "Prodigious quantities of honey are found here," observed a writer in 1773, "of which they make excellent spirits, and mead as good as Malaga sack."

Over-all Aspects of Colonial Agriculture

It is apparent, from all accounts, that agriculture in North Carolina was predominantly small-scale in character—where the major emphasis was placed on the production of a great variety of subsistence crops, rather than large-scale with an emphasis upon producing either tobacco or some other "money" crop. The number of production units that were devoted to a money crop was relatively smaller in the colony than in either Virginia or South Carolina. Undoubtedly, the most reasonable explanation of this difference is to be found in the fact that most North Carolinians had very poor access to markets. The colony did not have a single good port or suitable facilities for internal transport. North Carolina produced surpluses of a greater variety of agricultural products than either Virginia or South Carolina, yet, in the total value of her exports she was far surpassed by each. In fact, the volume of shipments from North Carolina ports near the end of the Colonial period exceeded only those from New Hampshire and New Jersey.

By present-day standards, the entire population of North

Carolina was rural. This means that the producer of a surplus of any kind had no local market, with a sizable body of discriminating buyers, in which to sell his product. The competition was between sellers rather than buyers, with the result that the local prices of farm products were usually very low, and most surpluses were disposed of by barter to a small local merchant in exchange for his wares. Very little cash was available or needed.

Under the prevailing circumstances, there can be little wonder that agriculture in Colonial North Carolina became increasingly "unprogressive." The incentives to excellence of performance were not present. It might be said, too, that the seeming abundance of land encouraged the occupants to become neglectful and indolent in their agricultural practices. These bad habits had a retarding impact on all phases of social development, with the result that North Carolina acquired a bad reputation among the thirteen states. Elkanah Watson, referred to above, probably voiced the prevailing sentiment concerning the state when he wrote, in 1786: "Perhaps no state had at that period performed so little to promote the cause of education, science, and arts, as North Carolina. . . . The lower classes of that region were then in a condition of great mental degradation."

Effects of the Revolution

Although the peace and quiet of the North Carolina countryside was disturbed by the Revolutionary War, the character of agriculture carried on in the state was but little affected by that conflict. As previously noted, the major objective in the agriculture of Colonial North Carolina was production of the necessary food supply rather than production of commercial crops. This objective was attained by the farmers in the state during the war. Undoubtedly there was some suffering for want of food in the areas in which military operations were conducted, and in those communities most sorely beset by the civil strife; but, in the main, the food supply was sufficient. In fact, in the latter stages of the war, North Carolina was a principal source of livestock and livestock products for both sides. A scarcity of salt, which was the most important item that was in short supply during the war, led to an increase in the practice of driving livestock to out-of-state markets.

The disruption of commerce occasioned by the war had serious consequences for the relatively small number of producers for the export trade of such items as tobacco, rice, provision stores, and naval stores. The British blockade was so effective that it was almost impossible to deliver those products to market. As a result, the imports of sugar, rum, and molasses, which had previously come from the British West Indies, were cut off pending the establishment of new trade connections. The short-

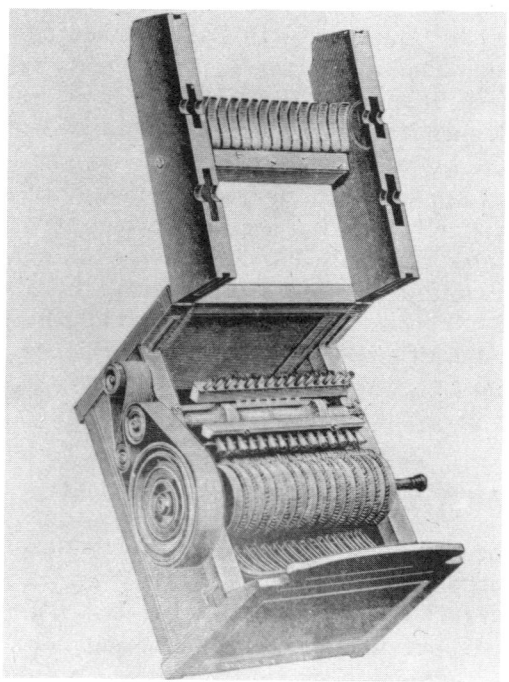

Model of the cotton gin. From Butterworth, *The Growth of Industrial Art*.

age of rum stimulated the building of many small distilleries in the state for the manufacture of whiskies and brandies. The practice of marketing surpluses of grain and fruit in liquid form became widespread in North Carolina, and distilling became one of the leading industries in the state for several decades. The inability to import textile goods during the war also encouraged the planting of more cotton. The product, however, was used in home manufacture, and did not become important in the export trade until after Whitney invented the cotton gin in 1792. By

14

that date, the volume of agricultural exports from North Carolina had recovered to pre-war levels. Yet, in dollar value, the total was greatly exceeded by that in both South Carolina and Virginia.

Conditions Following the Revolution

The conditions which explain or contributed toward the inefficient and unprogressive character of agriculture in Colonial North Carolina were not eliminated with the winning of political independence from the mother country. Down to the Civil

Archibald D. Murphey, advocate of reform and progress in North Carolina. Files of State Department of Archives and History.

War, the population continued to be overwhelmingly rural and devoted to agriculture as a means of livelihood. Since there was no sizable local market for his produce, the farmer never received the stimulus to improve the quality of his products that a large mass of discriminating buyers provides. The lack of a good port, and of satisfactory internal transport facilities, served to discourage production for markets outside the state. These handicaps to the improvement of agriculture were recognized by the more intelligent and progressive citizens of the state, but, despite the efforts made, they were only partially eliminated prior to the Civil War. Private initiative and resources for the removal of these deficiencies were insufficient. The deep-seated hatred of taxation in all forms precluded any action being taken by the state—even if it had wanted to act—until the last half of the 1830's when the state began encouraging the building of railroads.

North Carolina had no free, tax-supported public schools prior to 1840. As a result, the very high rate of illiteracy which existed in the adult population served to encourage the continuance of superstitious practices in agriculture, and to foster the development of unprogressive attitudes which served as obstacles to all efforts at reform. Under such circumstances, there is little wonder that North Carolina acquired a reputation as the "Rip Van Winkle" state in the union. William R. Davie, one of the founders of the University of North Carolina, wrote in 1805 that "the friends of science in the other States regard the people of North Carolina as a sort of semi-barbarians, among whom neither learning, virtue, nor men of science possess any estimation." Archibald DeBow Murphey, one of the most intelligent and progressive citizens of the state, observed in 1819: "I had no Idea that we had such a poor, ignorant, squalid Population, as I have seen . . . the Mass of the Common People in the Country are lazy, sickly, poor, dirty and ignorant."

Weighted down by ignorance, superstitions, and devotion to inefficient methods, the mass of farmers in North Carolina continued to practice a pioneer system of agriculture down to the Civil War. This system was described by Charles Fisher to the Rowan County Agricultural Society, in 1821, as follows:

> We pursue a course of agriculture that takes all from the earth, and returns nothing to it: We go on, year after year, tilling our fields, without any pains to return to the earth the strength that each crop takes from it. We completely exhaust our soil by an unvaried succession of

crops; **and,** when it **can** produce no longer, we turn it out into fields, let it **wash into gullies, and** grow up with pines, and broom sedge, that never failing **symptom of exhaustion.** This is the common fate of our fields; the system that is defacing our country, and ruining our lands.

George W. Jeffreys, one of the state's earliest advocates of agricultural reform, wrote in 1820: "Our present, is a land-killing system, which must be altered for the better; for if persevered in, it must ultimately issue in want, misery and depopulation."

Westward Migration

Undoubtedly, soil exhaustion and a scarcity of "new ground" were major factors in shaping the decisions of larger and larger numbers of North Carolinians to abandon the state and move

Notice of sale of farm and household property by Adlai Ewing, 1816, preparatory to move to Kentucky. Ewing farm located near Statesville. Files of Department of Archives and History.

westward. This emigration, which deprived the state of people from all levels of society—especially the young folk, began in the 1780's. By the 1820's it had reached such proportions that there was scarcely an issue of a local paper that did not carry an article describing a departure from the community for the West, or an advertisement of a farm for sale by a person contemplating such a move. With some emigrants, this urge to move west assumed the aspects of a panic in that they abandoned their landed property in North Carolina after failing in their efforts to find buyers. Archibald D. Murphey estimated that the state lost more than 200,000 people by emigration between the years 1790 and 1816. Despite the efforts made to encourage people to remain in the state, the losses continued high until the decade of the 1850's. Murphey observed that North Carolina had not cultivated a respect for itself, but had remained "careless and spiritless," and had, in effect, driven a large portion of its most useful population from the state by failing to encourage industry and enterprise. ". . . the true cause," he wrote, "is to be found in the want of public spirit, of State pride, and of State feeling."

From whatever causes the desire to emigrate sprang, the effects were to be seen on every hand. The price of land and improvements thereon, which was the state's principal form of wealth, was depressed far below its real value or replacement cost. The human loss, which was heaviest in the young middle-aged groups, including slaves, made up a sizable portion of the most vigorous and venturesome element in the population. Among those who left the state were three young men who later held the highest elective office in the nation: Andrew Jackson, James K. Polk, and Andrew Johnson. Quite understandably, a chronic state of low morale afflicted many who remained. Also, the loss of fluid capital, carried away by emigrants to the West, served to retard all projects aimed at the advancement of North Carolina.

Agricultural Revolution

Despite the many obstacles that stood in the way of progress, the effects of the Agricultural Revolution, which was brought to America in the 1780's, were soon felt in North Carolina. Launched initially in England by practical agriculturists, the efficiency of their farming operations and the quality of their

Bill of December 10, 1807, showing sale of wheat, linseed, cotton, and oats by William McClelland to Donaldson and McMillan at Fayetteville. Salt, sugar, coffee, molasses, and cash for difference in value received in return. Files of State Department of Archives and History.

products soon showed such marked improvement that progressive-minded farmers everywhere were encouraged to examine critically their own agricultural practices. Fortunately for the future of the state, a small number of North Carolinians participated in this movement for the reform of agriculture from the very beginning. Through personal contacts and correspondence, and the pages of local papers, these leaders sought to stimulate an interest in a more efficient agriculture. They were in the main, practical men—uninformed as to the scientific aspects of agriculture—but determined to improve the efficiency of their operations, and to put farming on a more "business-like" basis. For the first time they began to keep records of their activities in which notations were made of methods used in preparation of the soil, of seed selection, of time of planting, of culture in growth, and of all other factors thought to have any influence on the success of the crop. The results of these "experiments" were shared with others—either by conversation with neighbors, or through letters to the editors of the local newspapers. Gradually the number of "improving" farmers increased in the state as the leaders in the reform movement succeeded in stimulating others to follow their examples. Once the desire for improvement had been aroused, farming thereafter ceased to be a "dull, clodhopping business," and every aspect of agriculture excited efforts aimed at greater efficiency. The great majority of farmers, however, persisted in their outmoded practices, seemingly unaware that a reform spirit was in the air. This ages-old failure of man to profit from good example led "A Virginia Farmer" to raise the question, which to him was unanswerable, as to why "bad" farmers who had the example of the "good" before them did not reform. ". . . but they don't do it," he said, "and if we may judge the future by the past, they never will." The more discerning advocates of reform, however, recognized that time and the utmost of patience were essential before an agricultural revolution could be brought to pass. "While aiming at much," wrote the editor of a farm magazine, "we must be encouraged with little, and labor for more with patience, diligence and perseverance—having our purposes strengthened with the sentiment, that

> Not to go back is somewhat to advance
> Men must learn to walk at least before they dance."

Publicity for Agricultural Reform

The increasing interest in agricultural reform led several North Carolina newspapers, notably the Raleigh *Star*, to include sections on "Rural Economy," "Agriculture," and "Rural Affairs," in which a great variety of farm topics were discussed either by local correspondents, or in articles reprinted from other papers. Although ill-advised on occasion in their choice of articles for reprinting, and in placing an overly-enthusiastic stamp of approval on untested implements and practices just because they were new, the editors of such papers rendered an increasingly valuable service to the cause of agricultural reform down to the Civil War.

The almanac, perhaps, proved to be an even more important source of general information to the great mass of farmers than the newspaper or any other publication. Undoubtedly, it was more widely read in that day than any publication other than the Bible. Designed primarily as a record of astronomical data, the almanac also carried a wide variety of anecdotes, essays, cures, recipes, and articles on agriculture. Many people who had nothing but scorn for "book farming" found it within their conscience to use information obtained from an almanac.

For those North Carolinians who felt no prejudice against "book farming," two books made significant contributions to the cause of agricultural reform. The *Arator*, written by John Taylor, of Caroline County, Virginia, and first published in book form in 1812, became "an immediate sensation." Although uninformed as to the chemistry of soils, and the scientific processes involved in plant and animal growth, Taylor placed great emphasis upon the fundamental facts in agriculture, particularly the maintenance of soil fertility by a generous use of manures, both vegetable and animal, in planting. By 1818 Taylor's book was in its sixth edition.

One of Taylor's disciples, George W. Jeffreys of North Carolina, brought out a volume in 1819 entitled *Essays on Agriculture*, in which he echoed Taylor's criticisms and proposals for reform and strongly advocated such "First Principles of Agriculture" as:

> Industry and attention to agricultural pursuits, and intelligence therein, are indispensable to insure success . . . experienced and skilful farmers have in all ages discovered the necessity and utility of draining wet lands. . . .

One of the most important principles in agriculture is cleanly farming. . . . Every field in cultivation should be kept entirely clean.

Manures will always fail in producing the desired effect, in proportion as draining and cleaning are neglected.

A change, or a judicious rotation of crops is necessary, in order to keep the soil in good heart and to enable it to produce its utmost.

Selecting and propagating . . . the most approved kinds of grain and seeds, is the surest method of preserving them in perfection. Seeds should be selected in the same manner that breeders are selected.

Liberality in procuring good tools or implements for the hands on a farm is the economy of agriculture.

Foresight is another item in the economy of agriculture. It consists in preparing work for all weather, and doing all work in proper weather, and at proper times.

When a thing is done, let it be well done, and it will not require to be done soon again. . . .

The growing interest throughout the nation in the improvement of farming led to the publication of journals, or magazines, which were devoted almost exclusively to agricultural topics. The *American Farmer,* published at Baltimore, the *Southern Agriculturist,* published at Charleston, and the *Farmer's Register,* published at Shellbanks and Petersburg, had most subscribers among North Carolina's "improving" farmers. The success of these journals in stimulating the farmer's interest in reform led to the publication of several similar, but less well-known, journals in North Carolina. The titles, places, and dates of publications, and the editors of these journals were as follows:

The Farmer's Advocate and Miscellaneous Reporter. Jamestown (1838-1842). John Sherwood.
North Carolina Farmer. Raleigh (1845-1849). Thomas J. Lemay.
Farmer's Journal. Bath and Raleigh (1852-1854). Dr. John F. Tompkins.
Carolina Cultivator. Raleigh (1855-1857). William D. Cooke and Benjamin S. Hedrick.
Arator. Raleigh (1855-1857). Thomas J. Lemay.
North Carolina Planter. Raleigh (1858-1861). James M. Jordan, John W. Woodfin, S. W. Westbrook, and W. H. Hamilton.
Edgecombe Farm Journal. Tarboro (1860-1861). William B. Smith.

Although zealous in their efforts to promote needed reforms in all aspects of agriculture, not one of these journals attracted sufficient subscriber support to continue publication longer than four years! The president of the state agricultural society attributed this lack of support to a want of state pride. Dr. John F. Tompkins, editor of the *Farmer's Journal,* which was an excellent publication, commented that many of the farmers were so prejudiced and opposed to innovations that they considered everything projected for their benefit "a 'Yankee humbug,' and thus they repulse us, thinking our intention is rather to injure than benefit them." Generally, the editors emphasized the same ideas that Jeffreys called "First Principles of Agriculture."

Agricultural Societies

Another manifestation of the interest in reform may be seen in the founding of societies dedicated to the promotion of "all improvements in agriculture or in any species of domestic economy." The Cape Fear Agricultural Society, founded in Wilmington in 1810, was the first such organization in North Carolina. The editor of the Raleigh *Star*, in commenting upon this development, began recommending that similar societies be formed in every county in the state. Of course no such general response was made, but several societies were formed in North Carolina in the next few years. Perhaps the Rowan County society was one of the most active of these. Organized under a constitution that provided for two meetings a year, and the sponsorship of an annual fair, the Rowan society included in its membership "a number of respectable planters of the county." Each member of the society was pledged

> diligently to turn his attention (as much as his situation will allow) to the study of agriculture, and on all occasions to impart to the society any improvements or discoveries he may make; also, to use every exertion in his power to procure correct models of the most approved farming implements in use in any part of the county.

The first agricultural fair held in North Carolina was sponsored by the Rowan County society and held in Salisbury in 1821. The exhibits included a wide variety of vegetables, livestock, and agricultural implements. One of the animals exhibited was a large English Short Horn bull—to the owner of which an "honorary premium" was given "for his public spirit in bringing an animal of such superior quality into the county." Silver cups were awarded for the "most approved practical Essay on the subject of manures, particularly vegetable and atmospheric," and, an essay on the subject of raising livestock "in which the errors and defects of the present practices must be plainly pointed out and a better plan recommended."

The broadening interest in the promotion of reform led to the organization of the North Carolina Agricultural Society in 1818. This body held annual meetings in Raleigh for a number of years thereafter, and sought to encourage reform by urging the formation of societies in each county in the state by offering awards for prize-winning essays on selected topics, and by publishing those essays thought to be worthy of such honor. The direction that reform should take, in the opinion of the state society, was indicated in the following titles that were offered in the essay competition: "The Best Method of Reclaiming Worn-out Lands" and "The Production and Application of Manures Using Native Materials for Native Crops."

In 1822 the state legislature appropriated $10,000 to divide among the counties, in proportion to their federal numbers, for the purpose of encouraging agriculture and domestic manufacturing. To participate in this fund, a county was required to form an agricultural society, to raise a sum equal to the amount to be received from the state, and to send in annual reports of its activities. The act of appropriation also provided that the president of the state society, and a delegation chosen by that body should comprise the "Board of Agriculture" with the duty to receive and evaluate reports from county societies. The Board of Agriculture, after its organization in 1823, published several volumes on agricultural subjects, distributed literature, seeds, plants, grape vines, and silk worm eggs through the county organizations.

Despite the constructive work done by these societies, and the continuing need for organizations of this sort, they began to fall into decline and inactivity in the late 1820's. The same thing was happening elsewhere in the nation. Several explanations may be made for this development. Undoubtedly, the decline of agricultural prices after the Panic of 1819, the lack of markets for surpluses, poor transportation facilities, and the continuing lure of the West were factors contributing to the decline. Then, too, the organization and conduct of meetings of the typical society were such as to disappoint and discourage the "average" farmer. Too often the average farmer, for whatever reason, played the role of a mere listener.

Renewed Interest in Reform, 1845-1860

Despite the decline of interest, the impulse to organize for the promotion of agricultural reform did not die, and, in fact revived and expanded with increased vigor from about 1845 onward to the outbreak of the Civil War. Many factors and developments contributed to that revival in North Carolina and the nation: widespread improvements being made in transportation, communication, and marketing facilities; publication in 1840 of Justus Liebig's pioneer work on the chemistry of soils; improved livestock and farm implements more generally available; an increase in reading by the general public; the organization of the National Agricultural Society in 1841; and the influence of the first state fair in New York in 1841. Of particular importance for North Carolina, the tide of emigration from the state had passed its peak in the 1830's; the state entered upon a trend toward general progress after 1835; and more people recognized the need for agricultural reform.

With the revival of interest in reform, the number of county, community, and regional agricultural societies grew steadily in North Carolina after 1845. Again the state sought to encourage this development by offering an annual subsidy to each society to help defray the cost of awards to winners of contests it had sponsored. To receive this aid, each participating society was required to publish a report of its experiments and improvements in at least one newspaper. The organization, programs, and objectives of these new or revived societies were the same as in the earlier period. Unfortunately, the same deficiencies and failures were also present. The stiff formality with which meetings were conducted repelled the "average" farmer. He either did not join the society—and this was the course of the great majority—or he played a mute role in its activities. In commenting upon this, the editor of the first farm journal published in North Carolina said that most farmers were ignorant of accomplishments being made in agriculture. "And this is not all," he continued, "they manifest a determination to remain ignorant. . . ."

Organization of a New Agricultural Society

Chiefly as a result of the initiative taken by Dr. John F. Tompkins, editor of the *Farmer's Journal*—then being published at Bath—a meeting was held in Raleigh in 1852 in which the North Carolina Agricultural Society was organized. In taking this action, no note was made of the fact that an organization bearing this name had been formed in the earlier period. Delegations from eighteen counties attended the organizational meeting. John S. Dancy of Edgecombe County was elected president, a memorial was prepared in which state aid was requested to carry out the society's program, and plans were made to sponsor a state fair the following year.

The new North Carolina Agricultural Society attracted to its membership many of the most distinguished leaders in the economic, political, and social life of the state. Prominent among these were the four men who served as presidents of the society

Silver medal awarded by the North Carolina Agricultural Society, 1853. Files of State Department of Archives and History.

to 1860: John S. Dancy, Richard H. Smith of Halifax County, Judge Thomas Ruffin of Alamance County, and Dr. William R. Holt of Davidson County. These leaders, assisted by equally progressive-minded and practical advisers, helped the society to play an increasingly significant role in the progress of agriculture until the Civil War. Farmers were urged to subscribe to a farm journal; a "corn sweepstakes" competition was sponsored in which a prize was offered for the largest per acre yield of corn; and sizable annual premiums were offered for essays on a wide variety of agricultural topics. The best of these essays were published either in a newspaper, a farm journal, or in the transactions of the society, and farmers were encouraged to maintain a file of such articles for reference purposes in seeking solutions to their own problems. Undoubtedly, many new practices and ideas in agriculture, as well as implements, seed varieties, and breeds of livestock became known through the publication of these essays.

The state society sponsored a fair at Raleigh each year from 1852 until the outbreak of the Civil War. This project attracted the support and participation of a larger number of people each year, and, by 1860, was an event of state-wide significance.

Thomas Ruffin, one of the presidents of the North Carolina Agricultural Society, member of North Carolina Supreme Court. Files of Department of Archives and History.

The cash premium list broadened each year, and served to promote a lively competition among exhibitors. Beginning in 1855 the state legislature appropriated funds to assist the society in offering more attractive cash awards. Several of the railroads then serving Raleigh also gave the fair a boost by offering to transport exhibits to and from Raleigh free, and passengers for only half-fare. With these encouragements, the exhibits of plants, animals, farm implements, and products of domestic science grew in number each year. Undoubtedly, many farmers, who were passive in their reaction to other aspects of the movement for agricultural reform, found inspiration at the fair to improve their farming practices. No doubt, too, the carnival atmosphere that came to be an important aspect of the fair, provided the farmer and his family many new and sometimes pleasant experiences. "Let it no longer be said," the editor of the *Farmer's Journal* wrote, "that the farmers of North Carolina are blind to their interests, and that they have no spirit of pride as regards the advancement of farming among them."

"Scientific" Farming

By whatever means, whether through an agricultural fair, society, or the press, a principal objective of the agricultural reform program was to rid the farmer of his prejudices against new ways of doing his work, and to stimulate a "spirit of inquiry" concerning every aspect of farming. At about the same time that the editor of a Salisbury newspaper was raising the question: "Can Farming be an Intellectual Pursuit?" Justus Liebig, the famous German chemist, startled the agricultural world with his volume entitled, *Chemistry in Its Application to Agriculture and Physiology*. From the very beginning leaders in the Agricultural Revolution, although only amateur scientists, had been stressing that enrichment of the soil was a fundamental first step in the reform of agriculture. Liebig emphasized that substances necessary for the growth of a particular plant, and these could be identified by chemical analysis, had to be present in the soil in proper proportions to assure proper plant growth. This concept led to widespread experimentation and the gradual accumulation of "scientific" knowledge as to the properties and function of the soil. The North Carolina legislature reacted to these developments by authorizing the Governor, in 1851, to employ "some gentleman of science" to conduct an agricultural, geological, mineralogical, and botanical survey of the state. Professor Ebenezer Emmons of the University of North Carolina was employed as state geologist and placed in charge of the survey. The results were published in five volumes, between 1852 and 1860, and clearly illustrate the Liebig influence in the attention given to the chemistry of the soils of the state. A farmer wrote to the editor of the *Farmer's Journal* as follows: "Prof. Emmons is telling us where, and what our agricultural advantages and resources are; we look to you for instruction, as to the best means of developing and using them."

Fertilizers

As a result of the above developments, "improving" farmers gave far more attention to the application of manures and fertilizers in an effort to restore to the soil those elements now known to be essential to plant growth. Scarcely any substance was overlooked in the widespread search for manures to accomplish that purpose. The *American Farm Book,* published in 1850, gave an appraisal as to the chemical composition of each of the following substances—all of which were being experimented with by farmers in North Carolina: ashes, lime, marl, shell sand, green sand marl, gypsum, bones, phosphate of lime, salt, nitrate of potash and of soda, charcoal, guano, soot, barnyard manures, poudrette (human feces or "night soil" as it was called), blood, flesh, hair, fish, cotton seed, peat, seaweed, rich turf, swamp muck, and clover, cow peas, and other green crops.

Accounts of "experiments" with one or more of the above substances in the stepped-up search for increased soil fertility, occupied a larger and larger proportion of the space in agricultural journals as the Civil War approached. Such experimentation, of course, would have little chance of producing the desired results without adequate preparation of the soil to receive the fertilizer. Recognition of this fact greatly stimulated the desire for improved farm implements, and many old implements were improved and new ones introduced. This was the case with agriculture throughout the United States in that period. Never before in such a short period of time had so many changes been made in the mechanics of farming as were made between 1815 and 1860.

Farm Implements

The "improving" farmer in North Carolina was quick to avail himself of these new or improved implements, and some were ingenious enough to contribute toward the progress along that line. In the decade of the 1850's alone, North Carolinians

Wheel cultivator, patented in 1846. From Butterworth, *The Growth of Industrial Art.*

patented the following farm implements: 1 corn husker, 1 corn harvester, 3 cottonseed planters, 1 cotton-thinning plow, 1 cottonseed cleaning machine, 1 stalk cutter, 4 straw cutters, 7 plows, 10 cultivators, 2 seed planters, 1 hillside plow, 3 sowing machines, 1 harvesting machine, 1 thresher, 1 potato-planting plow, 1 wagon brake, 4 manure carts, 1 feeding apparatus for a grain thresher, 1 self-dumping truck, 1 marl lifting machine, 1 excavator, 1 hand truck, 1 water wheel, 1 pump, 1 carriage spring, 1 hoof paring knife, and 1 machine for stuffing horse collars. In the same decade, in anticipation of the better life, one North Carolinian patented a "self-waiting" table, and another a "ventilating" rocking chair. Impressive as the above list is, there is little doubt that many improvements were made in farm implements, particularly by local blacksmiths, that were not patented or produced for sale outside the immediate community. Perhaps the most effective promotion of the new implements was accomplished by the agricultural societies in their offers of premiums for the improvement of old implements and the invention of new ones. To win such a premium, the usual requirement was that the implements' merit had to be demonstrated at the fair.

To meet the growing demand for more and better farm implements, several concerns were established in the state which specialized in the production of farm implements of nearly every sort. A machine shop was established in Charlotte which began advertising that it could make anything "from a steam engine to a horse shoe." A foundry, located in Warrenton, began turning out forty plows a week. The Clarendon Iron Works, in Wilmington, began manufacturing farm machinery of many sorts. The state fair of 1860 exhibited cultivators, plows, corn shellers, seed sowers, horsepower machines, threshers, separating and cleansing machines, and tobacco presses and fixtures made by Frerchs and Raeders in their foundry in Salisbury. Similar businesses were located in Fayetteville, Tarboro, and Greensboro. Out-of-state manufacturers distributed their implements in North Carolina either through agents located at various places in the state or through sale of patent rights to local producers.

From the above it would appear that farmers in North Carolina, by the outbreak of the Civil War, were adequately supplied with the most up-to-date equipment. Unfortunately, such was not the case. The supply of good sturdy implements from all sources was never adequate. The forces of prejudice against innovations and the attachment to old habits were so strong with many that they persisted in using the old devices even though the superiority of the new was amply demonstrated. Such failure to pursue one's own best interest, for whatever cause, posed the most formidable obstacle to the reform of all aspects of agriculture. To those reformers who were discouraged by such failure, Ebenezer Emmons recommended "the utmost patience."

Commercial and Subsistence Crops

It has been noted earlier that from the very beginning of settlement in North Carolina the emphasis was placed upon the production of subsistence crops rather than upon commer-

Cotton growth and harvest. From Butterworth, *The Growth of Industrial Art.*

Transportation of cotton to market. From Butterworth, *The Growth of Industrial Art.*

Use of the saw gin, 1794 invention. From Butterworth, *The Growth of Industrial Art.*

cial crops. In other words, the agriculturists in North Carolina were overwhelmingly engaged in "farming" rather than in "planting." This does not mean, however, that production of the commercial crops was of little importance in the economy of the state, but simply that their production was relatively less important than in the typical southern state at that time. Geography offers a partial explanation of that fact—North Carolina was along the southern fringe of the tobacco belt of that day, and along the northern fringe of the belt in which cotton, rice, and indigo were produced. At no time was sugar cane grown, for commercial purposes, in the state.

Cotton

Cotton, for those who could produce it, possessed several advantages as a money crop. Very little skill and no expensive equipment were required in its culture. It was well suited to cultivation by slave labor in that in its culture year-around employment was provided for both the slave and his family. The product, too, was nonperishable, suffered little from abuse in handling from the field to the market place, possessed great value per unit of weight, offered large and small scale producers almost equal advantages in its cultivation, and found a ready sale. In view of these facts, there is little wonder that after Eli Whitney invented the gin, cotton soon "became a considerable object of culture" in North Carolina. Previously the crop had been grown widely in the state on a small scale for home use. After its culture on a large scale had become practical and profitable, North Carolinians joined others in the South in, as they expressed it in that day, "the cry about cotton."

No reliable statistics are available as to the size of North Carolina's cotton crops in the years prior to 1840. The trend of production, however, seems to have been steadily upward until about 1825—after which time many producers sharply reduced the acreage devoted to the crop with the result that very little cotton was produced in the state until after 1840. The explanation for this sag in interest in cotton as a money crop is not entirely clear. By 1825 the northern geographic limits of the area in which cotton could be successfully grown had been determined. Some growers, in abandoning the crop, said that the growing season was too short, or that the spring was "too backward," or that the soil was too thin. Other factors that undoubtedly influenced that action were that the price of cotton had gone down; the costs of marketing the crop continued high; the state was losing many of its people to the western movement in the 1820's and 1830's; and, by 1825, it had become apparent that North Carolina could not compete on even terms with the states of the Lower South in the production of the crop.

Beginning about 1840 North Carolinians began planting cotton again as a money crop. Some of the handicaps, noted above, had been at least partially eliminated by that time, and producers began using the more "advanced" practices in their cultivation of the crop. The more important of these were proper drainage of the soil; selection of seed; use of improved implements; adequate preparation of the soil; timely culture of the crop while in growth; and a generous application of manures, including Peruvian guano, to the soil. Despite this revival of interest in the planting of cotton in North Carolina, the percentage of the total crop of the southern states that was produced in the state declined each year to 1860. Comparatively, North Carolina was not keeping pace with the other cotton-producing states even though the number of bales produced increased each year. Cotton, despite the above, was the most important money crop produced in North Carolina in 1860, and, as such, had first call on the best lands, equipment, and labor force.

Tobacco

Tobacco, the leading money crop of Colonial North Carolina, also experienced its ups and downs in the pre-Civil War period. With the winning of independence from England, the American producer lost his monopoly in the empire market, and thereafter found himself in competition with producers in other nations. Then, too, many nations began imposing very high duties on tobacco imports. Once established as a part of the national fiscal systems of those nations, all efforts to have the duties modified or repealed proved unavailing. In fact, the tendency was to increase rather than decrease the rates. This, of course, had the effect of lessening the demand for American-produced tobacco, which like cotton sold on the world market, and of keeping the price depressed. Other factors contributed toward keeping the tobacco producer in a state of depression: Tobacco quickly exhausted the fertility of the soil; some planters shifted from planting tobacco to planting cotton instead; and the area in the United States in which the crop was grown was greatly expanded.

The tendency of the tobacco planter to move west to the counties along the Virginia boundary had begun in the Colonial period. These counties, particularly Caswell, Granville, Person, Warren, and Rockingham, lay beyond the fringe of the cotton-growing region; and tobacco producers there marketed their crops in Petersburg, Farmville, and Clarksville in Virginia.

The methods used in tobacco growing changed less than in the growth of any of the money crops, and the crop continued to require more labor per acre, much of it painstaking hand labor, than any other crop. This is one reason why advocates of reform in agriculture quite often urged tobacco growers to abandon the crop for some other less demanding crop. John Taylor wrote:

> It would startle even an old planter, to see an exact account of the labor devoured by an acre of tobacco, and the preparation of the crop for market. . . . He would be astonished to discover how often he had passed over the land, and the tobacco, through his hands in fallowing, hilling, cutting off hills, planting, replantings, toppings, suckerings, weedings, cuttings, picking up, removing out of the ground by hand, hanging, striking, stripping, stemming, and prizing, and that the same labor devoted to almost any other employment, would have produced a better return than tobacco. . . .

This survey did not include reference to the labor required in the preparation, planting, and care of the tobacco seed bed.

As was the case with cotton, several factors contributed toward a considerable increase in the amount of tobacco grown in North Carolina in the two decades prior to the Civil War. Significant among these were that a new flue-curing process— using charcoal—was introduced; better means of transportation were becoming available; prices inclined upward after 1840; and a new type of leaf was being produced. The latter development came to pass almost by accident in 1839, when the Slade brothers of Caswell County produced "bright" tobacco by planting a crop on thin, sandy, ridgelands—a type of soil previously thought to be practically worthless and most unsuited to the culture of the crop. When cured, this tobacco had a beautiful lemon-yellow color, was light and dry, and well suited for use as wrappers, or as "fine cut" for pipe and cigarette smoking. Culture of this type of tobacco, which almost immediately commanded premium prices in the market, soon spread to all the counties in the northern Piedmont section, and into adjacent areas in Virginia. Stimulated by these developments, tobacco-growing was on the rise in North Carolina in 1860, and the state ranked sixth in the union in volume of product.

Corn

Production of one or more of the cereal crops is of primary importance to any farmer whose aim is to be self-sufficient. Since this was the objective of the large majority of farmers in North Carolina, one or more of these crops was grown all over the state. Even the planter of money crops recognized the advisability of producing his own requirements of the cereals—he would have been indeed unwise to have neglected to do so. Of the various cereals, corn was by far the most important, and the most valuable crop grown in the state. Unlike several of the other crops, corn could be grown in every part of the state. Because of its acceptability as a food for both man and beast, its hardiness as a plant, and its dependability, there is little wonder that corn was without challenge for the honor of being the most "popular" crop grown in the state. Although most of the crop was consumed at home, there were planters along the Roanoke River and Albemarle Sound who produced corn for sale by the shipload in Norfolk, Charleston, Baltimore, Philadelphia, New York City, and Providence.

Methods used in the culture of corn varied more widely, perhaps, than in the culture of any other crop. This fact contributed toward the submission of numerous articles to the agricultural papers, and to frequent discussions of the subject in meetings of agricultural societies. Fortunately, corn almost never failed to return a crop even when slovenly methods were used in its culture. Although some growers persisted to 1860 in measuring their crops by the number of hills planted, rather than by acres, by that date most corn was being planted in furrows or drills, and cultivated by animal-drawn implements. Of course, opinions varied as to the best methods to employ even in the same kind of soil, in breaking the soil, types and amounts of manures or fertilizers to be used, width of rows, whether to plant in beds—separated by water furrows—or in furrows prepared in flat lands, type of seed to plant, frequency and type of cultivation, and when to "lay the crop by." Only the more "advanced" planters, such as the Burgwyns of Northampton, George W. Jeffreys of Person, and Ebenezer Pettigrew and Josiah Collins of Tyrrell counties, made any attempt at a scientific rotation of crops including corn. Per-acre yields of corn, however, showed a steady increase for those who employed the improved methods in its culture.

Wheat

Wheat also came to be grown widely over the state, with heaviest production centered in the Piedmont section, and along the valley of the Roanoke River. Although a few wheat planters, such as the Burgwyn brothers of Northampton County rank with the largest and most progressive in the nation, the planting of wheat did not assume the importance for North Carolina planters that corn did, nor did its culture excite equal interest of reformers.

Except for those few who planted wheat for commercial purposes the methods employed in planting the crop changed but little after the Colonial period. "It is generally sown on corn land," wrote an Orange County farmer in 1850, "and put in in a slovenly manner. Some do better, sowing on fallow ground broken up late in summer, or early in autumn, and ploughed or harrowed in." Planting was usually done in the fall months, with very little if any manure being used, and no further attention was given the crop until harvesttime in June. Grown in this manner, there is little wonder that per-acre yields ranged from only four to ten bushels, and that wheat was not suited to cultivation, except in conjunction with other money crops, by slave-owning planters. Although the reaper and threshing machines came on the market in the 1830's, most North Carolina wheat continued to be harvested by either the sickle, scythe, or cradle, and to be threshed by either the flail or trampling under the feet of cattle. Despite the above, it is significant to note that the amount of wheat grown in the state doubled during the decade of the 1850's. More growers, too, were using improved implements and methods in its culture.

Rice

Unlike wheat and corn, rice could be grown only in the coastal region of the state—more particularly in the southeastern coun-

Threshing wheat by use of hand flails. From Butterworth, *The Growth of Industrial Art*.

Threshing wheat with horses. From Butterworth, *The Growth of Industrial Art*.

ties of Columbus, Duplin, Sampson, New Hanover, and Brunswick. Climatic and geographic limitations were such that rice growing never assumed the importance in the state that it did in South Carolina. In fact, in 1859—a year in which the state produced its largest crop of rice—six-sevenths of the crop was grown in Brunswick County. The necessity for irrigating the crop, and the unhealthiness of the area in which it was grown led many planters to abandon rice and turn to planting cotton, corn, or wheat.

Fruit

Although fruit growing attained only slight importance in the commercial life of the state, the conditions under which this activity was carried on also drew the attention of agricultural reformers. George W. Jeffreys commented as follows:

> The feelings of a lover of improvement can scarcely be expressed, on observing the almost universal inattention paid to the greater number of our orchards, and that people who go to a considerable expense in planting and establishing them, afterwards leave them to the rude hand of nature; as if the art and ingenuity of man availed nothing, or that they merited no further care. . . .

In like manner, other reformers drew attention to the carelessness and inattention with which most fruit growers tended their orchards, and they urged adoption of more intelligent methods. Growers were told that a half a dozen choice fruit trees, planted in proper soil and carefully attended, would produce more and better fruit than one hundred trees tended in the usual manner.

The increasing interest in the improved culture of fruit trees led to the establishment of the Lindley Nursery on Cane Creek in Chatham County in 1826. This was, perhaps, the first nursery established in North Carolina. The initial stock was secured from leading nurseries in the United States, and the trees were tested for actual bearing before being offered for sale through a descriptive catalog. Of peaches alone, 127 varieties were tested for bearing and were advertised for sale. This, and other nurseries founded before the Civil War, made available a wide variety of improved orchard stock at reasonable prices.

In 1856 and again in 1858 highest honors were won by apples exhibited at the National Pomological Convention in New York City by the West Green Pomological Gardens and Nurseries of

Greensboro. In the latter year, North Carolina apple growers played leading roles in founding the Southern Pomological Society at a meeting held in Charlotte. The *North Carolina Planter,* which began publication in Raleigh in 1858, reflected the growing interest in fruit raising by devoting a section in each issue to horticultural subjects.

Cultivation of grapes, whether for table use or for use in the making of wine, received more attention in the agricultural press, perhaps, than any other item of fruit. Unlike most of the fruits, several varieties of grapes were native of the state. Of the varieties that attained some prominence in the nation, the "Lincoln," "Isabella," "Catawba," "Hickman," and "Scuppernong" originated in North Carolina. Of these, the scuppernong, which thrives in the sandy soil of the Coastal Plain, was the most popular for eating and wine-making purposes. Sidney Weller of Halifax County, who developed "American Systems" of wine-making and grape culture, wrote numerous articles setting forth the superior qualities of the scuppernong. Hardy of growth, requiring little or no cultivation, less affected by insect enemies and diseases, the scuppernong, he emphasized, was a dependable bearer that age did not seem to impair.

Vegetables

The need for improvement in the variety and quality of vegetable production was not overlooked by those who advocated agricultural reform. Paul C. Cameron of Orange County urged: "Surround your dwellings by fruitful and well kept gardens. . . . No man is better entitled to all the good fruits of the earth, than he who tills it." The editor of the *Farmer's Journal,* published in Raleigh, wrote: "Pass through the garden once a day at least, give it an hour in the morning, and another in the evening, if possible; no part of the farm will pay you better than the garden crops." Such admonition was usually followed by advice as to the best types of soil and modes of culture of a wide variety of vegetables.

Seed of "improved" vegetables of all sorts were advertised in newspapers and farm journals along with advice as to how and when to plant. The sweet potato continued to be the most popular vegetable grown in North Carolina. Of the state's 86 counties in 1860, only 7 had a production of less than 10,000 bushels, and 40 counties produced more than 50,000 bushels each of sweet potatoes. Heaviest production occurred in the eastern counties where the potato was in much demand as an item in the diet.

Irish or white potatoes, although not grown on so large a scale, were also produced in every county in the state. Peas and beans, which had minor importance in the Colonial period as commercial crops, were widely produced for home consumption. The planting of peas was generally encouraged by agricultural reformers as a soil-building crop in rotation schemes. Turnips and artichokes were similarly promoted for this purpose and as a source of food for both man and beast.

Peanut or ground pea was one of the few vegetables introduced into production successfully which had not been grown in North Carolina in the Colonial period. Its potential value as a commercial crop was just becoming apparent when the Civil War intervened.

North Carolinians participated to some extent in all of the agricultural "crazes," "manias," or "fevers" that from time to time swept over the nation in the period prior to the Civil War. Each in its time became the object of wild speculative build-up, overly extravagant claims, intense excitement, disillusionment, and final collapse. The craze which excited most interest in North Carolina related to the planting of mulberry trees for the feeding of silk worms and the ultimate development of a silk industry. This project swept over the nation from the late 1820's until its resounding collapse in the early 1840's. Some few enterprising planters profited from it—chiefly by sale of stock for the production of mulberry trees, but one may doubt that any lasting beneficial results were obtained. The idea that some good may come from such experiences was expressed by one agricultural writer, as follows: "Improvements in agriculture seldom come as pure improvements. They come in epidemics, mixed with a great deal of nonsense, and with more or less of humbug, and often of rascality."

In summary it may be said that North Carolinians, in the absence of an effective demand for surpluses of fruits, vegetables, and other minor crops, directed their efforts toward producing their own requirements of these crops. All indications point to the fact that by 1860 this goal had been substantially accomplished. Improvements in the variety, quality, and quantity of fruits and vegetables had been made, and there was little evidence of scarcity of these products in the state.

Livestock

Under the conditions that prevailed during the Colonial period, North Carolina produced a surplus of livestock, particularly cattle and swine, for sale in the markets of the North or in the West Indies. By 1820 these conditions had become so altered or changed that the state was no longer self-sufficient in the production of livestock, but was importing part of its requirements of these products from other states. Explanation of that development may be found from the following facts: By 1820 settlement in the state had become so dispersed that large expanses of "common" grazing lands were no longer available. Under the pioneer-type husbandry practiced in the Colonial period, no provision was made for the shelter, upgrading by selective breeding, or winter feeding of livestock. The men who rounded them up, branded them, and drove them to market were mere drovers rather than herdsmen. When profits become unacceptably low, such operators simply shifted to some other type of employment. Then, too, many planters in their concentration on production of staple crops neglected to produce their own requirements of livestock. This was the case, even with draft animals. Some of the most intelligent planters in the state, despite the criticisms and admonitions of agricultural reformers, were importing mules from Kentucky as late as 1860.

In an effort to arouse an enlightened self-interest in this direction, one reformer wrote: "Pay less attention to the breeding of hounds and more to that of horses. The sleek fat beeves and golden rosy flavored butter of New York, are not the produce of broom grass and scrub pines." "Never keep a poor or malformed animal of any kind," advised another, "it is better to kill or give away such, than incur the expense of keeping them, and the risk of having their peculiar deformities communicated to the rest of your stock."

Among those who advocated reform of this aspect of agriculture, two schools of thought developed: There were those who labored for a complete revolution of the livestock industry by adoption of new breeds with special consideration being given to the purposes the animal was intended to serve. That was the approach suggested by George W. Jeffreys, and followed by Dr. William R. Holt, president of the North Carolina Agricultural Society in 1859. Holt imported thoroughbred Devons

A type of feeder used in exceptional cases where livestock was given care and protection. From Butterworth, *The Growth of Industrial Art.*

and Durhams for his Davidson County farm and won much acclaim when he exhibited these animals at the State Fair. Paul C. Cameron, of Orange County, was a leader among those who recommended a more gradual approach toward improvement. He emphasized that no "costly importations" should be brought into the state until farmers were prepared to furnish full supplies of proper food and "a more humane care." After that, he would gradually improve herds by crossbreeding the imported animals with "judicious selections" from the native stock.

Unfortunately, too few of the mass of farmers and planters in the state followed the advice of either group. Although some progress was made in improving the quality of livestock, and, equally important, more attention was given to the production of forage crops, yet, most North Carolina farmers in 1860 were managing their livestock in the same manner as did their pioneer forefathers. In commenting on this state of affairs before the state agricultural society in 1853, a member said:

> It is impossible to pass through the country in the spring without being pained to observe the cattle which have just achieved the enterprize of enduring the winter.
> Those which have survived, give unmistakable indications that their perils have been great, and the danger of starvation imminent.

It may be safely concluded that by 1860 North Carolina had only entered the beginning stage of over-all improvement in the quality of its livestock and of livestock management.

Farms in 1860

Approximately 70 per cent of the landowners in North Carolina in 1860 owned less than 100 acres each. On the other end of the scale, only 2 per cent of the landowners held acreages in excess of 500 acres. The average farm had 316 acres and had become smaller each year after 1850. The pattern fixed in the Colonial period that North Carolina should become predominantly a farming rather than a planting state, was followed to 1860. The Census of 1860 listed 121 planters and 85,198 farmers in the state.

Approximately 70 per cent of the landowners owned no slaves, and therefore did all of the work on their farms with the assistance of their families and occasional hired help. Over two-thirds of the slaveowners in the state owned less than ten slaves each. Six thousand four hundred and forty persons, by the Census of 1860, owned only one slave each. The average size of the slaveholding in that year was 9.5—having declined from 10.1 in 1850. It may be concluded that as the slaveholding era was nearing its end the institution was becoming more popular with the farmer class. Where the need for extra labor existed, the farmers preferred either to hire or buy a slave rather than employ free Negroes or white laborers. This sentiment, of course, impaired opportunity for advancement of both the free Negro and the white laborer.

In 1860 North Carolina had a total population of 992,622—an average of nineteen persons per square mile. In the absence of any sizable industrial establishment, this population was overwhelmingly rural in its composition and the average farmer lived in isolation from his neighbors. Given this setting, and, in the absence of a good port or of adequate internal transport facilities, it is no wonder that self-sufficiency, rather than production of money crops, was the primary goal toward which the average farmer devoted his efforts. The same condition also led North Carolina planters to diversify their efforts rather than concentrate on the production of a single money crop. The over-all result was that the per capita value of food crops produced in the state in 1860 exceeded the average per capita production for the United States and that of each of her immediate neighbors. The diversity of the state's commitment to the production of money crops was reflected in Governor William A. Graham's comment, in 1845, that it would be difficult to say which of the state's export crops was of greatest value: cotton, tobacco, rice, wheat and flour, or corn.

By 1860 many of the circumstances that had served as obstacles to the improvement of agriculture in North Carolina were in process of being changed or modified. A public school system had been established which gave promise that the pall of ignorance would be gradually lifted from the people. Railroads and plank roads were opening up the interior of the state and providing, for the first time, access to markets to many citizens. A new surge of interest was manifested in the formation of agricultural societies on the local and state level, with programs aimed at over-all improvement in agricultural practices. Less ridicule was heaped upon "book farmers," and more attention was given to the scientific aspects of agriculture.

In remarking on these developments, Governor Thomas Bragg said in 1856, "Our greatest interest, agriculture, to say nothing of the others, is attracting the notice it deserves, and our people are on the inquiry for the best modes of improving their lands and increasing their crops." The editor of the *Farmer's Journal* commented that North Carolina "has at last aroused from her slumbers, shaken off her dull feelings and entered with her sister states, fully and fairly upon a thorough improvement in agriculture." Undoubtedly, the coming of the Civil War terminated, for a time, a trend toward over-all improvement in agricultural practices that gave promise of more prosperous times for all the people of the state.

SUGGESTIONS FOR FURTHER READING

Bennett, Hugh Hammond. *The Soils and Agriculture of the Southern States.* New York: Macmillan Company, 1921.

Brooks, Jerome E. *Green Leaf and Gold: Tobacco in North Carolina.* Raleigh: State Department of Archives and History, 1962.

Carman, Harry James (ed.), *American Husbandry.* New York: Columbia University Press, 1939.

Carrier, Lyman. *The Beginnings of Agriculture in America* (First edition). New York: McGraw-Hill Book Company, Inc., 1923.

Cathey, Cornelius Oliver. *Agricultural Developments in North Carolina, 1733-1860* Chapel Hill: University of North Carolina Press, 1956.

Gray, Lewis Cecil. *History of Agriculture in the Southern United States to 1860.* 2 volumes. Washington: Carnegie Institution of Washington, 1933.

Hobbs, Samuel Huntington, Junior. *North Carolina: Economic and Social.* Chapel Hill: University of North Carolina Press, 1930.

Johnson, Guion Griffis. *Ante-Bellum North Carolina: A Social History.* Chapel Hill: University of North Carolina, 1937.

Lefler, Hugh Talmage, and Albert Ray Newsome. *North Carolina: The History of a Southern State.* Chapel Hill: University of North Carolina Press, c. 1954.

Neely, Wayne Caldwell. *The Agricultural Fair.* New York: Columbia University Press, 1935.